ABOUT *HUGGING THIS ROCK*

"Whether running down Minnesota trails with his beloved dog Leo; flying the skies above Baghdad and Kandahar; or navigating the frontiers of mid-life, marriage, and parenthood; Eric Chandler takes on the natural world and the human condition with grace, determination, and a wicked sense of humor."

—*Randy Brown, author of the poetry collection* Welcome to FOB Haiku

"Eric Chandler's pilot perspective allows for attention to detail—the strength behind a handmade sawhorse, the terrain of mountains seen from above, the measures parents take to protect their children, a pre-deployment bar conversation—while simultaneously seeing the broad landscape of identity."

—*Lisa Stice, author of the poetry collection* Uniform

"Chandler's verse transmits imagery with clarity rivaling northern Minnesota's lakes. Layering combat pilot's-eye views on endurance & escape; family & landscape; war & what comes after; it provokes & refreshes.

I devour his poems with gusto usually reserved for a handful of gooseberries or black caps, washed down with a long drink served by the cold pump handle at the cabin."

—*William Schuth, poetry editor,* The Deadly Writers Patrol

"It's always a pleasure when a poet takes the world you know and hands it back to you at a new tilt. [...] Chandler's lens brings us our familiar world from thousands of feet above and then zooms into close range, so that [...] we can feel what matters far more keenly than we could on our own."

—*Andria Williams, author of the novel* The Longest Night

"Eric Chandler delivers a payload of philosophy, poetry, and memoir, while on a mission to answer the question of what makes us each more than the air we breathe."

—*Jason Poudrier, author of the poetry collection* Red Fields

"This is stealth poetry. With each poem, Chandler glides you gently along, then, *BOOM!* He hits when you least expect it."

—*Susanne Aspley, author of the novel* Granola, MN

Other works by Eric Chandler

Outside Duluth – essays on nature,
recreation and sport

Down In It – a war-fiction novella

■■■

Other works published by Middle West Press LLC

Welcome to FOB Haiku:
War Poetry from Inside the Wire,
by Randy Brown, a.k.a. "Charlie Sherpa"

Reporting for Duty: U.S. Citizen-Soldier Journalism
from the Afghan Surge, 2010-2011,
edited by Randy Brown

Hugging This Rock

Eric Chandler

Middle West Press LLC
Johnston, Iowa

■■■

Poetry / Subjects & Themes / Places / Minnesota
Poetry / Subjects & Themes / War / 21st century

Chandler, Eric
Hugging This Rock / Eric Chandler
ISBN (print): 978-9969317-4-8
ISBN (e-book): 978-9969317-5-5
Library of Congress Control Number: 2017954837

The text of this book is set in Garamond Premier Pro.

■■■

Middle West Press LLC
P.O. Box 31099
Johnston, Iowa 50131-9428

www.middlewestpress.com

Cover image: Senior Airman Donald Acton,
Minnesota Air National Guard

Author photo: Staff Sgt. Chris Axelson,
Minnesota Air National Guard

To Dorothy Chandler

CONTENTS

EARTH

Ditches

I feel like we are
rolling in riches
violet lupines
fill up the ditches

Official Flower

Two years ago,
I ran along the Superior Hiking Trail
with my dog.
We saw a violently violet flower
right next to the constant hum of
an interstate highway.
Neither of us knew what it was.

My daughter sent me a photo
with her phone this week.
She told me there are ten
blue flag irises
in our garden this year.
She reported this is better
than the two we had last year.

Last year, she told me that
the blue flag iris
is her favorite flower.
A flower with such blue that
her blue-gray irises are only distant
color cousins that
encircle her pupils
and let in the light.

We have lived in
this Duluth house for fifteen years.
The irises have always been there.
My daughter is twelve.
The irises have always been there.
We ran to the flower with no name.

The irises have always been there.
I didn't plant them.
I didn't see them.
Decades passed until
I saw what was
in my own garden.

Pay attention.
Fifty years to find
a blue flag iris.

Pay attention.
Fifty years to find
your favorite.

Now,
Quebec and I
both have
an official flower.

Mother's Day

Vancouver, Canada
May 8, 2016
7.7 miles

1.
An observation platform
Next to the nest where
The swan preens
Unperturbed

2.
The needle sharp shadow of
The heron strikes
In the shallows

3.
The woman bikes by
Her screaming kid
In a trailer
"Look, buddy!
A big rock!
With a tree on top!"
Good try, Mom
Good try

4.
"All these things that happen to us are
Irreversible."

5.
A woman plucks a koto
In the sunshine
So improbable

6.
Another heron
Perfectly still
Hundreds pass and I can see that
I'm the only one who looks

7.
"... how good it is to live here."
"You are lucky."

Retention Time

The urban legend:
All the cells in your body
are replaced every seven years.

The water in me must turnover
faster than seven years.
A spoonful of water would have a retention time of
maybe just a couple of days.
A week? A month?
I cross-country ski and
my sweat
pours
out.

I squint at the snow
in the late winter sun.
I imagine it melting and
flowing into Lake Superior
past the steelhead and
out the flood-changed mouths of the
North Shore streams.
A drop of snowmelt taking
two hundred years
to make it to the Soo.

I Won't Remember

I won't remember this
late spring pre-dawn run.
Record snows melting in the ditches.
The deer tracks in the snow.
The harsh call of a red-winged blackbird.
The flight of two honking Canada geese.
The familiar awakening of my limbs as I pass two miles.
I walk across the surface of an untracked field of snow
and the entire slab settles with a whoomp.
There's nothing remarkable about the full moon
as it sets to make room for the rising sun.
So, I won't remember this morning in North Dakota.
It's likely that,
at some point,
I won't remember anything at all.

Moving Stuff Around

At first, you only move the air in and out.
But then, you move milk.
The milk changes and you move it into your diapers.
You move some blocks on top of other blocks.
You move your bicycle up and down the street.
You move the clothes you wear to school and back.
You move them into the washer and into the dryer
and back onto your body for a road trip.
You move the petroleum into the tank
and you move tons of metal and rubber and glass
down the highway.
You're lucky and
you get a job as a train engineer or a truck driver or an airline pilot.
You're really moving stuff around now.
Millions of tons that you move and move and move.
You move her body.
Her body moves iron and oxygen and nutrients into the new part
of her body.
You move your kids into the car and move them to piano and karate
and to school.
Eventually, they move away to another town where
they keep moving food onto plates and plates into sinks
and onto racks and onto shelves and
back onto the counter where they
move more food onto them.
Later, you move the spoon.
The spoon helps you move the canned soup
into your face so you can move the broth
into your kidneys and out in a halting stream.
At the end, you only move the air in and out.
You move the oxygen in and carbon dioxide out.

Maybe you move the air or maybe a machine helps you.
Maybe the air is moving you.
Maybe you're being breathed.
Maybe the atmosphere is moving your ribcage in and out.
They move you into the ground or an incinerator.
Your calories move into the dirt or the sky.
Some people will move water out of their eyes.

Anger Management

The Zen monk says,
You are crossing a river
by ferry at night.

A man
in a boat
in the dark
is coming straight for you.

You yell
at the man
in the boat
in the dark:
Turn away,
damn you.

The man
in the boat
in the dark
hits you.

Then you see
the boat is empty.
Your anger evaporates.

Even if there was
a man
in the boat
in the dark,
do not be angry.
He knows not what he does.

Just like
an empty,
rudderless
boat.

Sure, but
that man
in the boat
in the dark
has free will.

And he keeps screwing up the river.

The Old Man of the Mountain

For decades,
I flew every sortie
with a New Hampshire quarter
in my pocket.
The quarter with the Old Man of the Mountain on it.
It kept me safe.

Then, in 2003,
the Great Stone Face fell.
It was in the middle of the night.
There was a loud noise and
the Old Man of the Mountain,
who held vigil over Franconia Notch for
ten thousand years,
fell off the side of Cannon Mountain.

My lucky quarter didn't seem so lucky.
But I still flew with it every time in Afghanistan.
When I got back home,
I stopped flying with it.

My Uncle died recently.
I went to his funeral.
I drove home with my parents and
had too many whiskeys with my Dad.
In the middle of the night,
exactly ten years after the Old Man fell down,
I got out of bed to piss.
When I reached to flush the toilet,
the lever didn't work.

I woke up drooling on the carpet.
My arm hurt.
My back hurt.
I fell down, apparently.

My dad came downstairs.
He said he heard a loud noise.

I haven't felt quite so close to
Live Free or Die
in a while.

Hetch Hetchy

There are two signs on
The towel rack.
One says, "cozy" and explains that
The towel rack
Heats your towels.

It's next to the switch
That fires up
The electricity to the towel rack.
That fires up
The coal fired power plant.

The power plant
Sends up the gas.
Is the drought because the power plant
Sends up the gas?
Either way, there's a drought.

I looked down through that gas at the
Hetch Hetchy reservoir.
White bathtub rings surround the low
Hetch Hetchy reservoir
Because of the drought.

The second sign on
The towel rack
Says they won't launder what's on
The towel rack.
Only what they find on the floor.

All the water in the city comes from
The Hetch Hetchy.
They're conserving water from
The Hetch Hetchy.
They hope you won't mind.

Enjoy your hot towels.

Zeitgeber

The shower tells me it's morning.
The bowl of cereal tells me it's 0630.
The coffee in my truck
tells me it's 6:40 a.m.
The traffic light tells me
it's 11 minutes before the hour.
The loon chicks tell me it's June.
The fireflies tell me it's July.
The fireweed tells me it's August.
That smell.
That smell
in September tells me
winter is coming
as the fireweed
blooms
to
the
top.
I don't want the shower to
tell me it's six in the morning.
I want the shower to
give me more time.

Popocatepetl

(via "Romance" by W. J. Turner)

She said the name playfully.
Popocatepetl.
I've seen that volcano
covered with snow
in Mexico.

I stood where Popocatepetl
In the sunlight gleams.

The graffiti reeled past
the windows of the van.
Coloring the cinderblock square
and filthy air.
But just up there ...

Shining Popocatepetl
The dusty streets did rule.

She stirred the pot
and called me by my sister's name.
Her mind was getting worse.
It was a curse.
But she still quoted verse.

She sang ditties and
poems to the kids.
She mimicked the chickadee-dee-dee
and laughed with glee
with my sister and me.

We sat by her bed and
saw the clubs of her hands.
I watched her eyes close.
Her chest fell and rose.
I wonder if she knows?

After a decade, could we set her free?
Her son said: It's not up to me.
Maybe in there she can see
the chickadee.
The chickadee-dee-dee.

O shining Popocatepetl
It was thy magic hour:

The houses, people, traffic seemed
Thin fading dreams by day;
Chimborazo, Cotopaxi,
They had stolen my soul away!

I couldn't watch the decay and
fled to the sunroom.
The TV evangelist felt the need to scold
the gathered, gray lumps. The old.
I ran out into the cold.

Maybe, behind her eyes, she said the name playfully.
Popocatepetl.
There's no way to know
if she saw the snow
on our volcano.

Labels

The wind blows and
The aspens say
Ssshhhhh.
The red pines say
Fffffffff.

The shower hits.
The poplars say
Ssshhhhh.
Norway pines say
Fffffffff.

Aspens or
Poplars.
Red pines or
Norway pines.
Different labels.
The same kind of trees.
Whatever they're called,
They answer the breeze.

Crossing the T

I tumble down the hill
like water in a creek bed.
Left and right ...

Finding my way home
through the city blocks
like river rocks.
Finally.
There it is.
At the bottom of the street.

Sometimes, the sun blinds
as it bounces off.
Or the steam rises at
twenty below zero.
Sometimes it's gunmetal gray.
Or brown and green.
Even a turquoise ad for a
vacation in Jamaica.
There are whitecaps with wind
from the east.

A thousand-foot laker
just might inch past
the end of the street—
Like someone
crossing a giant T.

Sawhorse

I almost killed Phil more than once.
I backed out of my driveway and braked for the
Green Bay Packers jacket and white cane
taking Tuffy the West Highland terrier for a walk
in the dark during a snowstorm.
It was all the same to blind Phil.
He messed around across the alley
in his garage with his power tools
while his wife Marty tended the flower gardens and
the giant rhubarb plant.
He made me a pair of sawhorses.
Two-by-fours with remnants of red and green paint
held together with long deck screws.
Still amazed he didn't cut off
one of his fingers in the
permanent dark.

Not sure what to do with them,
I took them to the cabin.
They ended up holding the boatlift
above the frozen water.
Eight legs immobilized in the white
with their heavy load.
In the same spot when the spring waves
lapped at the now-gray wood.
Sentinels in the ice,
year after year.
Indestructible.

I visited Phil when his insides failed.
I saw him hold his hand in the air.
Kind of waving it, never able to see me,
—not before, not now—
from that bed in hospice.
Marty, said, "He wants to hold your hand."
I should've known something was up when, earlier,
he gave my kids all his fishing poles
and me all his bottles of brandy.
I see Marty walk with Tuffy every day.
Badasses out in all the snow and ice
that Duluth delivers,
just like Phil.

Last year, one of the sawhorses was swept away.
Just disappeared.
What good is one sawhorse?
I went to the cabin with two new plastic ones.
They joined the boatlift team with the sole survivor.
I stomped through the knee-deep snow this winter to check on things.
The plastic swaybacked under the load while
Phil's sawhorse stood stoic surveying the snow.
Bearing the weight like Atlas.

Three Inches or Smaller

Pull the rope
and the two-stroke engine
spins the chain.

Knock down everything
three inches in diameter
or smaller
that juts
into the trail.

The chain pulls
the bar through the tree.
The flesh of the balsam
invites its killer
deep inside.
The Old Man taught me to
let the saw do the work.

Later,
as I glide
over the snow,
the branches will not slap my face.
Will not brush my shoulder.

The gasoline
and the oil
and the fir
dance together.
There will be more trees.
They can't be stopped.
I can't be stopped.

There will be no smell later.
Just cold air
in and out.
You will never know the saw was there.
You will never know I was there.

The Wave

You're at the game and
It starts somewhere
Over there and you can hear
The noise and the raised hands
And the roar gets closer
And closer
And closer
And it's almost your turn
And Holy Crap
YOU'RE STANDING UP
WAVING AND SHOUTING
And then you sit down
Laughing as the sound recedes.
It's exciting.
You're in the ground and
It continues somewhere
Over there and you can hear
The life and the warm hands
And the pulse gets closer
And closer
And closer
And it's almost your turn
And Holy Lord
YOU'RE STANDING UP
RAVING AND SHOUTING
And they lower you down
Laughing as the sound recedes.
Was it exciting?

Speed Limit

From the enclosed cab
of my truck.
Catkins.
Trillium.
Bunchberry.
Red pine pollen.
Dandelions.
Dandelions in my lawn.
I'm a dandelion farmer.
I'm a rhubarb farmer, too.
Fluffy cottonwood seeds.
collecting on the edges of the driveway.
Lupine in the ditches.
Peonies under the sign at the cabin.
Orange Hawkweed
(a.k.a. Indian Paintbrush).
The tamaracks are green again
in the swamps.
Columbine.
Heals All.
Black Medic.
Daisies, plain old daisies.
When the fireweed
blooms
all the way
to
the
top,
it means there are
six more weeks
until winter.

It's passing
faster than the speed limit.

Tan Hermosa

The bike path is green
like part of a LEGO set.
The salty air and skateboards.
The beach volleyball games.
(It's Tuesday.)
The man with aviator sunglasses
riding too fast
on his beach cruiser.
Flip flops on the pedals.
The tires roll through the dog shit
and leave echoes of the mess
with each revolution
until the sand and sun
dry-clean the wheel.
Two lions guard the entrance
to a beach house:
Oceanfront Viagra.
The fat kid buzzes by
on a battery-powered scooter.
Strollers
and walkers
and runners
all playing their music on speakers.
I'm meant to share, I guess.

I veer to the pier.
I can finally hear the surf.
Morgan Freeman reminds me that
this ocean has no memory.
I lean against the railing.
The Asian dude casts and jigs.

My bad eyes
splinter the sunspots
into triangles.
It's quiet even with a breeze.
I look at the surfers.
Here comes a set.
And, just for a second,
I'm cool with it.

SKY

View-Master

When you pull down on the lever,
the cliffs of Crete
dive into the Mediterranean.
A new picture snaps into place
and you cross the Nile
for the ninth time.
You see how small
Pearl Harbor is tucked into
the ocean with no memory.

When you pull down on the lever,
you go five hundred miles per hour
over the mile-wide frozen Yukon River.
A new picture snaps into place
and the glowing Northern Lights
twist and flick over Fairbanks.
You see the Apostles
swimming in Superior
like giant green amoebas.

When you pull down on the lever,
the asteroid streaks across the sky and
breaks into three pieces.
A new picture snaps into place
and the Hindu Kush reaches up to
pull down the sky.
You see the man
walk his camel down a road
just over the border from Turkmenistan.

When you pull down on the lever,
the surface of the Salt Lake
mimics the clouds.
A new picture snaps into place
and you look down at Fremont Peak
where you stood with your friend.
You see the glittering arc
of mortar shells over the Tigris
into Sadr City at night.

When you pull down on the lever,
you kick the rudder and stare back
at the giant triangle of Denali gilded by the sun.
A new picture snaps into place
and the Rub' al Khali
stretches off into a beige forever.
You see the airspeed indicator
almost hit twice the speed of sound,
but not quite.

When you pull down on the lever,
the Rock of Gibraltar juts
as you glide toward the sensuous Spanish hills.
A new picture snaps into place
and the White Mountains emerge from shadow
and glow orange in the rising sun.
You saw so much
because the number of landings in that machine
matched the number of takeoffs.

When you pull down on the lever,
you appreciate the poplars of Kyrgyzstan
almost as much as those in Duluth.

Hugging This Rock

We're both down here hugging this rock but when I fly I go faster which means I'm getting younger than you because Einstein's special theory of relativity says that as I go faster the velocity dilates time and it slows down in relation to those on the surface of the rock (which for your sake would be nice because your misery would pass quicker and my flying would last longer) and I'm grateful because I've seen a bunch of cool things up there but unfortunately there's a price because as I get farther from the rock and experience less gravity Einstein's general theory of relativity says that my time is actually going faster than the people that are driving down the highway to jobs in cubicles on the rock and I wonder whether I'm younger from flinging myself through the sky at high speed or maybe I'm older because I was vertically separated from my home for three hundred days but I may never know the answer to that question (since I'm not a scientist) although I suppose I could Google some more and figure out which effect is greater and I'd be able to tell whether all this flying gave me more time down here but either way I've seen a bunch of cool things up there that I can think about while we're both down here hugging this rock.

Lightning in my Chest

It was a Thunderbolt.
A plastic model.
We put it together.
We painted invasion stripes
on the outside after it was all built.

It was *Baa Baa Black Sheep*.
A television show.
We watched it together.
The gull-winged Corsairs
peeled off for the attack.

It was a Warhawk.
A real-life warbird.
We watched it start from behind the fence.
The paddles started to spin and it came
right for us until he kicked the rudder.

It was a P-38 Lightning.
At that time, supposedly only eight left.
We watched it at Oshkosh.
The gleaming, unpainted twin booms glided
and the engines purred.

We left the airshow.
An Avenger soared
as we walked to the car.

We drove around Lake Michigan
through the U.P. to get home.
I waded in that lake and thought about planes.

I went to Saudi Arabia and Iraq and Afghanistan.
I thought about the sun
glinting off those twin booms.

In my chest, I can still feel those engines.
The Lightning propellers.
Humming.

Looking Up

I've known many dogs but this one is strange.
When he hears a jet, he looks at the sky.
In the world of dogs, he has a sixth sense.
Other hounds hear sounds but don't wonder why.

We ski on the trails out by the airport.
He sees Vipers fly and starts to give chase.
He runs right for them, tearing through the woods.
Maybe one of these days he'll win that race.

I told a friend my dog chases aircraft.
He said he knew a border collie once
that did the same thing. It's making sense now.
My dog's half that breed and also half dunce.

It's easy to find pilots in a crowd.
If planes fly by, they're the ones looking up.
Fate must've brought me this weird shelter dog.
Odd little black-and-white plane-obsessed pup.

So, at a picnic with folks all around
with the kids and the dogs all running by,
if a Beaver on floats flies overhead,
Leo and I will look up at the sky.

Nobody Will Play It

the pilots took off to defend the island
not all of them came home
one of the pilots played the piano for the missing
a last tribute

when the piano player
didn't come home from his mission
his mates pulled the piano into the yard
and set it on fire

all of the pilots
all of these flames
fly up into the air

if he can't play it
than nobody will play it

the fingers of my children
fly up and down the keyboard
they celebrate the arc
from takeoff until the fall to earth

I will pull the piano into the yard
and listen to the strings melt
and the hammers burn
and the red
and the smoke
and the flames
will be
my last tribute

all of these notes
all of these flames
fly up into the air

if they can't play it
then nobody will play it

The Program

As the First Officer, it's my job to throw the switch. This causes the chemicals to flow from their special vat into the fuel tanks where it mixes with the gas. The gas flows into the engines where it burns in the combustion chamber. The chemicals flow out with the exhaust and trickle down on the people. You can see the white trails in the sky behind my airplane. The chemicals induce obesity, selfishness, and whining. Results are much better than expected.

It's Only Light Chop

Hey, citizens.
Are you feeling okay?
I only ask because
you keep hitting me in the shoulder
as I sit in the aisle seat.

He Smokes Cigarettes

He smokes cigarettes.
I don't like it one bit.
But less than he used to.
He's trying to quit.

He's funny, has kids,
knows his job really well.
Then he throws a tantrum.
I think: What the hell?

He lost fifty pounds.
He's improving his life.
He did this because he's
cheating on his wife.

It would be nice if
people made it easy
for me to know if they're
okay or sleazy.

No Big Deal

He comes over the P.A. and says there's a
"little problem with the landing gear."
This is my line of work,
so I'm curious.
He says everything's fine.
We flare and
wait for the rubber
to screech onto the pavement.

We didn't go cartwheeling into the ditch
like the six-million dollar man.
I thanked my fellow pilots.
And the tower for looking over the gear
during the go-around.
And the engineers who designed a system
that checks itself and
checks itself again.
The rubber screeched onto the pavement.
No screeching people.
And it's no big deal.
It's no big deal.

Man-Machine

The pilot who wrote *The Little Prince* said
we take the airplane for granted.
Because of its modern reliability,

"... the machine does not isolate man
from the great problems of nature but
plunges him more deeply into them."

In high school, my buddy Mel
had a thought about
the pads that football players wear:

They protect you so well that
if you even notice you've been hit,
it's already
really
really
bad.

One flier always shakes the crew chief's hand
before closing the bubble of glass and
fusing himself to the fuselage:

It might be
the last time
I touch
a human being.

Vainglory

the sun shining down from behind his head
the shadow of the plane lies on the cloud
circular glory from purple to red

the young pilot's ego needs to be fed
metal wings strain and the engine is loud
the sun shining down from behind his head

old pilots know what the scientists said
glories don't care what you've done to feel proud
circular glory from purple to red

the Brocken specter should fill him with dread
it gets bigger nearing the icy shroud
the sun shining down from behind his head

the science of light won't let glory spread
changing the glory's size is not allowed
circular glory from purple to red

shadows grow to meet you until you're dead
yet, the halo of light remains unbowed
the sun shining down from behind his head
circular glory from purple to red

LOVE

Prayer Wheel

When I can't be home,
I do this.
I go to bed and
say the name
of the city out loud.
"I'm in Billings."
Or, "I'm in Iraq."
"I'm in Afghanistan."
Or, "I'm in Newark."
It prevents
disorientation
when I wake up.
(Although today I thought
Orlando was Chicago.)

Then I roll to the right
and look where she should be.
I imagine all of us upstairs
in our beds
in our house
in different rooms
but so close
we can say goodnight
through the thin walls
like *The Waltons*.

My wife's feet point to the lake.
My son's feet point to Duluth.
My daughter's feet point to the North Star.
Heads to the center.
Toes out.
A pinwheel of people.

I exhale while I think of my wife's name.
Then my son's.
Then my daughter's.
And then I fall asleep.

Bubble

I am not there.
But I am listening.

My little one makes a joke
And your laughter runs down the stairs like a river.
She says it like a question
And you laugh and laugh.
And I would be able to see all your teeth.
My son joins in.
I would make jokes too
If I knew I would be showered with love.

I sit downstairs and lean back in my chair.
I see this bubble of joy in my house.
I am not there.
But I see it floating.
Upstairs.

You Are So Loved

Along the Lakewalk
in the sun.
It's probably 75 degrees.
Hot for me.
Just offshore, the fog bank
parallels the shoreline.
A laker silently approaches the canal,
barely visible in the leading edge of the mist.
Slips between the light and the silent foghorn.
There's no captain's salute.
The bridge has nothing to answer
so also stays quiet.
I run around the farthest corner
of this giant lake.
I can see that the bridge is down again.
I didn't hear the bell that goes with its descent.
Did the fogbank muffle all the sound?
Did a laker just go through the canal and
under the bridge behind that thin white veil?
I run from the clear hot sun into the fog.
It's chilly now, probably 55 degrees.
Two women are drawing with chalk on the pier.
Giant six-foot-tall letters that span the concrete
from one wall to the other.
It spells out:
You Are So Loved.
I run up the steps like Rocky,
round the lighthouse in the fog,
and head for home.

That Feeling

you know you're asleep
you know you have to wake up
you have to wake yourself up
you grunt and yell within the dream
you strain in that uncomfortable place
between sleep and consciousness

I hate that feeling
maybe that's why
ghosts are so angry

Buddhist, My Ass

If all things are one, then
it should also be true,
you feel all emotions
at once, too.

So, when they crack her skull
to rip the tumors out,
you should still find things to
laugh about.

But since you're a coward,
when they lower her down,
you'll weep. You'll cry. You'll feel
like you drown.

Common Enemy

Our half-breed dog. Leo's his name.
He comes over and watches the toast.
Circles the toast like a moth round a flame.
His yellow eye-lasers burn and roast.
You show him your hands and say, "All done."
He wags a little and walks away.
Without the toast, you're just no fun.
"Food motivation" is the phrase of the day.
Just before bed, he tries some affection.
The Labrador in him wants to come sit.
The Border Collie part goes the other direction.
Social. Anti-social. They don't really fit.
Oh look. A puddle. I think he just peed.
He's the common enemy you know we all need.

Epicanthus

my son is one of the mouse people
brown hair
brown eyes

my daughter has pale blue eyes
unlike her mother
or her father

if you look closely at their eyes
you can see
the epicanthal fold

the grandmother of my children
hid with her mother and sister and brother
to escape the Japanese

after the war, she left mainland China
and went to Taiwan
not fleeing communism, just finding work

she had a dozen jobs
to help her family
while she was still a teenager

my mother-in-law came to America
with my father-in-law
a few years earlier it would've been illegal

my mother-in-law raised three children
a daughter, a son
and another daughter

the in-laws all served in the military
a father, a daughter, a son,
and another daughter

and my mother-in-law
brought a genetic echo
across the Pacific Ocean

my wife looks Asian,
but don't get too close to my kids
do not discover the epicanthal fold

it will spoil the camouflage
I gave them

Bed Socks

Summer in Duluth.
The sun's finally here.
It's about dang time.
Winter hurt this year.

But change is so hard.
My wife is so sweet,
but she cannot stand
no socks on her feet.

I don't mean when she's
walking around town.
I'm talking about
when she goes to lie down.

Murmurs and mumbles.
What is she saying?
She's holding some socks.
Seems like she's praying.

"Do I dare try to
lay down my tired head
without my dear socks
when I go to bed?"

She gives it a try
with bare-naked feet.
But soon she sits up
and throws back the sheet.

Ten months in a row
She's covered her toes.
They cannot be free.
That's just how it goes.

She pulls on the socks
and she lies back down.
She's wearing a smile
instead of a frown.

I'm dripping with sweat
in the summer heat
but she needs bed socks
wrapped around her feet.

Youth

You have a face that does its best to
show emotion.
But you haven't had your heart ripped
out of your chest.
You don't really believe that there's evil
in the world.
You still think that the lady screaming
unashamed
at her kid in the checkout line at Walmart
is just a bad parent and not actually the manifestation of something
much
much
worse.

I Can Already Hear It

Afterwards,
the flags sprouted up like flowers,
an explosion of unity.
I was old enough to know
they would tatter and fade
like the public's passions.
The magnets that support the troops
would fall off the cars into the ditch.
A good thing would turn into a demonstration
of short attention spans.

I could already see it.

I sit at my breakfast table.
I read *National Geographic* and
crunch on some cereal.
I'm old enough to know
that the music drifting in from the other room
won't last.
My kids will grow up and leave home.
The piano that I don't know how to play
will fall silent.

I can already hear it.

It Happened Anyway

We bought fire extinguishers.
We changed the 9-volt batteries in the smoke alarms.
We even bought a rope ladder.
It happened anyway.

We ate lots of fruits and vegetables.
We exercised.
We ran marathons.
It happened anyway.

We told them not to talk to strangers.
We picked them up at school on time.
We helicoptered just like we were supposed to.
It happened anyway.

We always wore our seat belts.
We made sure they wore theirs.
We obeyed the speed limit.
It happened anyway.

We drank lots of green tea.
We did the crossword.
We challenged our minds.
It happened anyway.

We got a dog.
We made sure all the locks worked.
We installed an alarm system.
It happened anyway.

We taught them the meaning of truth.
We taught them to work hard.
To take responsibility. To be themselves.
It happened anyway.

I boiled the oil.
I barred the doors.
I begged.
And it happened anyway.

Quid Pro Quo

Something was wrong with her blood.
The minnow in her belly
might be growing incorrectly.
I got this news over the phone.
I was away from home.

Forty days in the desert on that trip.
I started running around the airport.
The sun was setting.
The heat rose from
where it was stored in the pavement.

I stopped next to the runway.
That was the one and only time
I have ever prayed.
If begging is praying.
I promised to teach my child about God.

My daughter is the only Minnesotan in my family.
She is not afraid of the cold.
She can cross-country ski faster than people twice her age.
She can paddle a canoe.
She caught a huge Northern in high style.
"Hold this," she said as she handed me the rod
and ran to get the net.
She knows how to find the North Star.
She can tell a white pine from a red pine.
I taught her these things.

I look into her healthy blue eyes
and ask her:
What are the only two things
that matter?

She points to her head and says:
What's in here.
Then she points to her heart and says:
And what's in here.

I will teach her more
if you need me to,
but I did my best
to hold up my end.
I wanted you to know I didn't forget.

WAR

You Should Know

It turns out,

I'm the kind who
says, "I told you so."

Before we start,
I thought you should know.

Air Born

a portajohn in Kyrgyzstan
one of my favorite pieces of graffiti:
Toodles, Afghanistan

our chartered airplane followed the great circle west
over territory I didn't recognize
a long sweep of coastline
probably the Maritimes
the sun gleamed down
through the severe clear

over the St. Croix
between Maine and Canada
reversing waterfalls are nearby
one direction when the tide flows
and the other when it ebbs
I knew where I was then

over the White Mountains
tiny from the air and
vast in my memory
my Limmer boots walked over
forty-eight peaks
I saw my birthplace

I tapped my friend on the shoulder
in the seat in front of me
I pointed down
with a war hangover and said
I was born there
Littleton

we slid across the northern tier
over the Upper Peninsula
over the old runways
he pointed down
I was born there
Kincheloe

faces plastered to the window over the
places we appeared by accident

Four Things

"Admit nothing.
Deny everything.
Act surprised.
Show concern."

Some people say you should
"Make counter accusations."
I didn't learn it that way.
Besides, you'll tip your hand.

For example, you would say,
"What's going on?
It wasn't me.
Really?
Oh, I'm so sorry."
You can almost always pull this off.

If you add,
"I think you were the one,"
you'll highlight yourself as
the guilty party.
The counter accusations
are just a little too much.

Just relax.
End by showing your concern.
Your humanity.
It's more believable that way.

New Kid's Lament

I don't understand
why you get extra credit
for your inertia.

Get Off My Lawn

They rang my doorbell
and ran.
After several weeks of it,
my daughter asked,
"Am I going to be abducted?"
That was that.

I snuck out in the dark
and caught them.
I told them to
stay out of my yard.
I put profanity mustard
on that sandwich.

They told me
their actual names.
They were too stupid to bolt.
Too stupid to lie.
Back in my day,
we would've done both.

What Do You Do?

Elbows on the bar
in hot southern Spain.
On the way to
embrace the pain.

I've got two kids.
His child is new.
"It's really tough.
What do you do?"

When I leave the house
and tell them goodbye,
I'm completely sure
I'm going to die.

I make my peace
and set them free.
It might not for you,
but it works for me.

Slipping the Surlies

(via "High Flight" by John Gillespie Magee, Jr.)

Oh! I've slipped the reflective belt and dirt,
And danced the skies on my dust-covered wings;
Sunward I've climbed, and tried to stay alert
With my go-pills and flew a thousand rings
You have not dreamed of—wheeled over the dung
High in light-brown violence. Capping there
I've chased the stinking wind along, and flung
My eager craft through sandstorms in the air ...

Up, up the long, dry, afterburning view
I've topped the piddle-pack with easy grace.
Which drinking a Rip-It will make you do—
And, while with silent, drifting mind I've trod
In a narrow altitude block of space,
—Put out my hand, and slewed the Sniper pod.

The Stars and Stripes is Free

The *Stars and Stripes* is free.
I grab one to read
while I eat my eggs.
They're okay but the
bacon's not quite right.

A girl's face filled with glee.
The photo on this
page is black and white.
Her mom says this is
a party for her.

She is about to see.
The next picture shows
tears where the smiles were.
I stop chewing my
breakfast and lean back.

She fights to break free.
Her mother holds her
down inside some shack.
The other woman
dives in with a knife.

Why won't they let her be?
Why are the black robes
threatening her life?
The Kurd women cut
off some private skin.

I sip my hot coffee.
The Kurds are my team
in this war I'm in.
I see my own girl
getting blood, not cake.

It's hard for me to see.
A lie so awful
it makes my heart break.
To see daughters bleed
while I eat my eggs.

Spark

There were three
planting an
I.E.D.

Two were dead.
The third man
ran and fled.

We could see
him running
to be free.

He fell down
in a ditch
near the town.

All alone
with us two
and the drone.

The night's dim,
but with gogs,
we see him.

A clear shot.
Couldn't miss.
The gun's hot.

Lawyers talked
too long. He
stood and walked.

They won't say
"Call in." He
got away.

He went to
the town. What
did he do?

To the first
house to tell
them the worst.

Then to see
the next guy's
family.

I saw then,
deep in a
crowd of men,

he was hot
from his run.
The bright dot.

His live spark
still showed up
in the dark.

"Your man's dead,"
was what he
must've said.

Would he ask
to be free
from that task?

Would he choose
to die or
tell that news?

To this day,
I don't know
what I'd say.

Maybe I Should've Lied

The teacher asked
me to come to the class
and talk about flying.
He was my son's teacher and
the jet's always popular.

How fast? How high?
Pretty standard stuff.
I wore my flight suit
and handed out stickers even
though they weren't toddlers.

One kid asked
if I killed anybody.
I was surprised and
shouldn't have been.
I told him the truth.

Later that day,
in the squadron,
I asked a buddy
what he would've done.
I would've lied, he said.

I answered the question
in front of my son.
The only time it has come up.
"That's what happens in combat."
Next question, please.

Did Joseph Heller Know King David?

God asks you to destroy your nation's enemies.
So, you pick up a sling
and kill the enemies of your people.
You get blood on your hands.

You eventually become king.
You get the people all fired up to build a temple to honor God.
You make the plans for the temple.
You gather the materials for the temple.
But then,
God says you don't get to build the temple.

Because you have blood on your hands.

Hold on, now.
Say that again?

Walking from Korea

it was spitting snow
the kind where
you don't need the wipers going full speed
but you just can't figure out
the right amount of wiping

he wore a jacket and a hat
I saw him everywhere in all seasons
year after year
sometimes those locations were miles apart
on the same day

I wanted to talk to him
but now it's too late

he walked and walked and walked
I knew I couldn't be the only one who noticed
people thought he was crazy and called it in
I thought he might not be
the police interviewed him and said he was fine

he was a medic in Korea
he wept in his mother's arms
when he returned
according to his sister
the newspaper said Changed by War

how would I talk to him
now it's too late

the cops looked out for him
he didn't want their help
his sister looked out for him until she died
one lady tried to corral him while the cops came
but they knew the deal

he injured his ankle once
when it healed
they put him in assisted living
he stepped into the room and said it was nice
grabbed his coat and walked out the door

I wanted to talk to him
but now it's too late

Frozen Chosin
Pork Chop Hill
Incheon
Pusan
Who knows

he picked up rubbish
along the road in his neighborhood
and walked
I kept seeing him
sometimes with his coat over his arm

what would I say to him
now it's too late

he walked and walked
the only sane one among us
doing real things
walking and picking up trash and
shoveling out bus stops and walking

the girl driving the car
was not charged with a crime
he died crossing the road
while he was walking
I don't remember if the weather was bad

I wanted to talk to him
but now it's too late

I learned all this about him
in the newspaper
after he died
the newspaper said that after the war
he lost his joy

he used to play the piano for hours
but then
he stopped playing
and stopped talking
and started walking

why would I talk to him
he didn't want to talk
I could've walked with him
but now it's too late

Where's the Kaboom?

Our hero
is in a place of comfort.
(I'm the hero.)
My wife and I want to live in a nice place and ski and make babies.
Horrible things happen and the plan goes in the shitter.
We enter an unfamiliar situation in a northern town.
We adapt to it.
"It" is a constant series of vacations to the Middle East.
We get what we want.
We live in a nice place and ski and make babies.
(Just not where we expected.)
We return to our familiar situation
in a place of comfort.
We have changed.

I pay a heavy price for it.
That's the missing step.
Did I pay?
Will I have to pay?
There was supposed to be an earth-shattering
kaboom.

ACKNOWLEDGEMENTS

"Air Born" appeared in *Line of Advance*, May 2017. (Winner of the 2017 Col. Darron L. Wright Award for Poetry)

"Anger Management" appeared in *The Thunderbird Review* Vol. 4, April 2016.

"Bed Socks" appeared in the *Duluth News Tribune*, Aug. 18, 2013.

"Bubble" appeared in *The Talking Stick,* Vol. 22, 2013. (Honorable Mention–Poetry)

"Crossing the T" appeared in *Grey Sparrow Journal,* January 2017.

"Did Joseph Heller Know King David?" appeared in *Line of Advance*, Vol. 4, May 2015.

"Four Things" appeared in *Grey Sparrow Journal*, Issue 24, Spring 2015.

"Get Off My Lawn" appeared in *The Talking Stick*, Vol. 25, 2016.

"Hetch Hetchy" appeared in *Aqueous Magazine,* Vol. 7, Winter Solstice 2014.

"Hugging This Rock" appeared in *Sleet Magazine,* Vol. 8, No. 1, Spring-Summer 2016.

"I Can Already Hear It" appeared in *PRØOF Magazine,* Vol. 1, Issue 3, July 2014.

"I Won't Remember" appeared in *Aqueous Magazine*, Vol. 3, Winter Solstice 2013.

"It Happened Anyway" appeared in *O-Dark-Thirty,* Sept. 17, 2015.

"Lightning in My Chest" appeared in *The Deadly Writers Patrol*, No. 12, Spring 2017.

"Maybe I Should've Lied" appeared in *Ash & Bones,* Memorial Day 2015.

"Moving Stuff Around" appeared in *Line of Advance*, Vol. 4, May 2015.

"Popocatepetl" appeared in *Grey Sparrow Journal*, Issue 26, Fall 2015.

"The Program" appeared in *Sleet Magazine,* Vol. 6, No. 2, Fall 2014.

"Quid Pro Quo" appeared in *Line of Advance,* Vol. 4, May 2015.

"Sawhorse" appeared in *Sleet Magazine*, Vol. 7, No. 2, Fall-Winter 2015.

"Slipping the Surlies" appeared in *Proud to Be: Writing by American Warriors,* Vol. 6, Southeastern Missouri State University Press, 2017.

"The Stars and Stripes is Free" appeared in *Line of Advance,* May 2016. (Winner of the inaugural Col. Darron L. Wright Award for Poetry)

"View-Master" appeared in *The Deadly Writers Patrol,* No. 12, Spring 2017.

"What Do You Do?" appeared in *Sleet Magazine,* Spring 2017.

"You Should Know" appeared in *O-Dark-Thirty,* Nov. 16, 2013.

"Youth" appeared in *Aqueous Magazine*, Vol. 3, Winter Solstice, December 2013.

NOTES

EARTH

"Zeitgeber": The word "zeitgeber" translates from German as "time giver." Zeitgebers are external cues that help your body clock know what time it is. Pilots routinely travel across multiple time zones. I use rituals like a shower and a cup of coffee to trick my body into believing that, yes, it's really morning here in Germany, even though my body thinks it's night in Duluth.

"Popocatepetl": My grandmother recited a little ditty with this word. It comes from two Nahuatl words: "popoca" meaning "smoking" and "tepetl" meaning "mountain." I think she pronounced it "popo-CATA-petl." I looked for "El Popo" while flying and saw it smoking to the southeast of Mexico City. Later, I disappeared down a web-searching rabbit hole and discovered a poem called "Romance" by W. J. Turner. The poem mentions the mountain standing out, pristine above the grimy streets—just as I had just witnessed riding in the hotel van delivering us to our layover accommodations. Turner's poem also mentions two additional volcanic mountains in Ecuador. I haven't been lucky enough to see those yet.

"Crossing the T": Duluth is a city built on a steep hill that rises from the shore of Lake Superior. The streets follow the contour lines "east" and "west" along the shoreline (in reality, the shoreline stretches northeast and southwest). The avenues go "north" and "south" right up the fall line of the hill. The poem locates itself at 26th Avenue East, as it tips down toward the water. The lake shimmers in the distance, just past the northern terminus of Interstate 35, and within view of the most scenic Holiday gas station in America.

Sometimes, a vessel is there at the bottom of the hill. "Lakers" are the biggest ships on the water. They only work the Great Lakes. The smaller, ocean-going "salties" come into the port via the St. Lawrence Seaway. It always makes an impression when I come down the hill and a laker's there. It feels lucky, too.

How often does this happen, I wonder. One out of 500 times? One out of a thousand?

SKY

"View-Master": This poem looks back on some of the most amazing things I've seen from an airplane, as a series of brief images. As a framing device, I though of the old stereoscopic View-Master toys, which looked a bit like binoculars. Using circular disc of photographic images, you could cycle through 3-D images by pulling down a trigger-like lever. Discs often presented scenes from around the world. The View-Master was the 1970s version of virtual reality for kids.

Places mentioned: The Apostle Islands National Park is clustered just off the shore of Wisconsin in Lake Superior. The Hindu Kush is the jagged mountain range that stretches from Afghanistan, east through Pakistan, to the Himalayas. Turkmenistan is the country that borders Afghanistan to the north. Fremont Peak is the third-highest mountain in Wyoming, located in the Wind River mountain range near Pinedale. Sadr City is a section of the city of Baghdad in Iraq, near where the Tigris River winds its way through town. Denali is the name for the mountain formerly known as Mt. McKinley in Alaska. The Rub' al Khali, the desert that fills much of the Saudi Arabian peninsula, is also called the Empty Quarter. The White Mountains are the mountains in northern New Hampshire. Manas Air Base in Kyrgyzstan was a pitstop where we paused after leaving Afghanistan. Before flying back to the United States from there, I went running through the trees, much like I do at home.

"Lightning in my Chest": This is a poem about how plastic airplane models, a TV show, and an airshow turned me into a pilot.

It's fun when I list off aircraft by just their name and not their number, but here's a little code-breaker for some of these World War II propeller aircraft that inspired me:

- **P-47 "Thunderbolt."** A fighter aircraft, sometimes called the "Jug" because of its burly shape. Invasion stripes were alternating black and white stripes applied to the wings around the time of D-Day, so that U.S. and allied aircraft could be easily identified and distinguished from the Germans.
- **F-4U "Corsair."** A bent-wing Navy and Marine aircraft, which featured wings that could be folded to make room during aircraft carrier operations. In my opinion, the Corsair was the true star of the 1976-1978 TV show *Baa Baa Black Sheep*, which told stories of the Marine Fighter Squadron 214. Actor Robert Conrad played Maj. Gregory "Pappy" Boyington.
- **P-40 "Warhawk."** An aircraft from the early years of the war. The nose was famously painted with shark teeth when the aircraft flew with Chennault's Flying Tigers in China. Oshkosh is the location of the famous Experimental Aircraft Association "AirVenture" annual airshow and fly-in.
- **P-38 "Lightning."** A twin-engine beauty flown by Richard I. Bong, whose hometown was Poplar, Wisconsin. At a count of 40, Bong holds the most aerial victories of any American pilot, and is known as "Ace of Aces." You can drive over the Richard I. Bong Memorial Bridge on U.S. Route 2 to get to Bong's hometown of Poplar, Wis. And, just across from Duluth via the John A. Blatnik Bridge, you can visit the Richard I. Bong Veterans Historical Center in Superior, Wis.

The P-38 is my all-time favorite World War II airplane.

- **TBF "Avenger."** This was another carrier-based Navy and Marine aircraft that delivered torpedoes. Former U.S. President George H.W. Bush flew one of these in World War II. It's big and ungainly, so, even as a kid, I was impressed with its aggressive acrobatics that day at Oshkosh.

We lived in Cadillac, Michigan at the time and it was a long drive over the top of Lake Michigan through the Upper Peninsula (the U.P.) to get home.

"Looking Up": "Viper" is the informal military nickname for the F-16 single-seat jet aircraft. If you ever hear someone refer to it unironically by the official name of "Fighting Falcon," you'll know they've never flown it and never been near anybody who has.

The DeHavilland "Beaver" is a high-wing, single-engine aircraft that's loved and renowned by bush pilots. It's thrilling to hear the loping sound of that radial engine. They're always beautiful and never more so when they're equipped with floats, to land and take-off on the water.

"The Program": A conspiracy theory oft-heard by pilots says the government is spraying mind-control chemicals on the public from commercial aircraft. Instead of properly referring to the white water-vapor trails behind airliners as "contrails," the loons refer to them as "chemtrails." I always want to ask these people, "The same government that can't get coordinated enough to fix potholes and bridges? That one?"

(Of course, my denial could be a further part of the vast chemtrail conspiracy. Ask yourself this, however: Are pilots really that clever?)

"It's Only Light Chop": "Light Chop" is the level of turbulence just one step above no turbulence at all. Sure, as a trained and experienced flying professional, I always assume that I can walk in an airplane in all weather conditions, as easily as a sailor aboard a ship on a stormy sea.

Usually, of course, I keep my seat belt on.

"Vainglory": In this poem, I talk about two weather phenomena that I often see from an airplane. They were first noticed by mountain climbers. One is called a "Brocken specter." The other is called a "glory."

The Brocken specter is the viewer's shadow as it appears on a cloud or in a fog bank. The Brocken is a mountain in Germany where this phenomenon was first noted. It's an eerie apparition, because as the cloud or fog dissipates or thickens, the viewer's shadow alternately retreats and advances, getting bigger as it approaches.

A glory is a visual effect caused by water and ice crystals in the air, and appears as a rainbow-colored halo. Just like the Brocken specter, it occurs in a line of sight directly away from the sun, as the bright circle of light is reflected back to the viewer. Although you often see the Brocken specter and the glory at the same time, the size of the glory always stays the same. In other words, as you approach a cloud in an airplane, the shadow gets larger, but the apparent diameter of the glory is fixed.

The poem is in the form of a villanelle. I wanted to try this 19-line form, because it seemed very restrictive. There are five three-line "tercets," followed by a four-line "quatrain." The first (A1) and third (A2) lines are rhymed, and are also each repeated as a refrain. The rhyme scheme can be notated as: A1,b,A2; a,b,A1; a,b,A2; a,b,A1; a,b,A2; a,b,A1,A2.

Simple, right?

I once listened to Sherman Alexie talk about poetic forms on a podcast once. I'm paraphrasing, but he said that poems were like dancing and that poetic forms were like a straitjacket. He said that if you could dance while wearing a straitjacket, you were one heck of a dancer. *Challenge accepted!*

By the way, my publisher notes the word "vainglory" is defined as "inordinate pride in oneself or one's achievements." I have no idea what he's talking about.

LOVE

"Prayer Wheel": *The Waltons* was a TV show back in the 1970s. It depicted the life of a family living in Appalachia during the Depression. They lived in a modest home that had very thin walls. At the end of each episode, the family would famously say goodnight to each other through the walls as the camera panned back from the house: "Good night, Ma." "Good night, John-Boy."

This note and the one about the View-Master make me feel old.

"Common Enemy": Leo is a good dog from Iowa that we rescued from the local animal shelter. He's got good points and bad points. The main bad point is that, even after puppyhood, he pees on the floor now and then. The working theory is that it happens when he's sleepy. (Scene: Greeting the dog at dawn. "Good morning, Leo." Leo pees on floor.) Leo is family, and occasionally shared hazards bring family closer together.

I looked up the rhyming scheme for an English sonnet and tried it. Later, I learned sonnets also use iambic pentameter. Way too hard. Maybe on the next one.

"Epicanthus": The epicanthal fold is a fold of skin on the upper eyelid that covers the inner corner of the eye. It's common in people with Asian heritage.

I wrote this when I was angry.

"Quid Pro Quo": In Latin, the title means "this for that" or "a favor granted in return for something." My plea felt desperate like a prayer offered up from a foxhole. It was self-serving, but sincere.

A "Northern" is short for a fish called a northern pike. I guess I'm not supposed to capitalize it. But I did anyway.

WAR

"Air Born": "Limmers" are hiking boots made by Peter Limmer & Sons, Intervale, N.H. They are the Rolls Royce of hiking boots. I've had a pair for over 35 years. I've had them resoled, but they're still kicking.

My dad had a pair of Limmers that he wore to work with the U.S. Forest Service. He took me to the Limmer shop around 1980. I got a pair of custom boots, special because the upper is made out of one single piece of leather with only one seam at the instep.

I mean, they weren't custom for *me*. After all, my parents were Maine Yankees (code for "frugal"). They were custom boots made for another person and returned. They were for sale on the "used" rack. I didn't care. They were Limmers and they fit me fine. I climbed all 48 of the Four-Thousand Footers in the White Mountains of New Hampshire in those boots. I've climbed peaks in Korea and Alaska in those boots. In 2015, I climbed a Fourteener in Colorado in those boots with my wife and two kids.

I love those boots as much as you can love inanimate objects.

"Slipping the Surlies": Flying is what John Gillespie Magee, Jr. is talking about in his poem "High Flight" when he says "Oh! I have slipped the surly bonds of earth." I've often used these words in conversation, as a light-hearted way to tell people that I'm flying. (You: "What're you doing this afternoon?" Me: "Slipping the surlies!") When I was a cadet 4th class at the U.S. Air Force Academy, Colorado Springs, Colo., I had to memorize "High Flight" and shout it on request for the upperclassmen.

There are many parodies, of course, but I haven't seen one for the Global War on Terror. Mine features the hated uniform reflective belt; the fatigue-mitigating dextroamphetamine go-pills; "moon dust"; the popular Rip-It energy drink; the targeting pod that I looked through on every sortie; and the little plastic "piddle-pack" that I used for restroom breaks in my single-seat cockpit.

In other words, it's got it all!

"The Stars and Stripes is Free": I'll admit that I'm trying to make you think of the Flag of the United States when you first read the title. In the poem, the *Stars and Stripes* is a military newspaper that is distributed by the Department of Defense. All four times I went to the Middle East after 9/11, they had a stack of these free newspapers sitting at the chow hall. It's a smaller, tabloid-sized newspaper, so it was handy to eat your chow and read the paper at the same time.

In case you didn't get it somehow, the newspaper story is about ritual female genital mutilation.

"Spark": An I.E.D. is an "Improved Explosive Device"—a land mine cobbled together from various parts, such as an unexploded artillery shell, triggered by an old washing machine timer or cell phone. They are often buried in roads, or set into walls and curbs.

"Did Joseph Heller Know King David?": In 1961, Joseph Heller published a famous satirical novel set in World War II called *Catch-22.* The novel is about fliers in a fictional unit of B-25 "Mitchell" bombers. Crew members don't have to fly the dangerous missions if they're crazy—but if they ask not to fly, they clearly value their own safety and must be sane. Thus, they have to go fly. The phrase "catch-22" has come to mean an impossible, contradictory situation. A vicious circle.

King David was the second king of the United Kingdom of Israel and Judah. As a young shepherd, he famously defeated the Philistine giant named Goliath in battle with a stone and a sling. David became king, conquered Jerusalem, and brought the Ark of the Covenant to the city. David wanted to build a temple to house the Ark, but in 1 Chronicles 22:8 (NIV) he says:

> *But this word of the LORD came to me: "You have shed much blood and have fought many wars. You are not to build a house for my Name, because you have shed much blood on the earth in my sight."*

THANKS

Thanks to my editor and publisher (and fellow poet!) Randy Brown for helping find an overall shape for this collection. While we've known each other for years on-line—he even has family in Duluth—we first met in person when I participated on a "Citizen-Soldier-Poet" panel he moderated at the 2017 Association of Writers and Writing Programs conference in Washington, D.C. Our discussion was subtitled "Using Poetry to Bridge the Civil-Military Gap." I'd like to think *Hugging This Rock* is a concrete example of doing just that.

Thanks to Veterans Writing Project founder Ron Capps and Managing Editor Jerri Bell for being the first to publish a "military" poem of mine at the organization's on-line and print journal *O-Dark-Thirty*. Ron's book *Writing War* and Jerri's insights have inspired me to create local opportunities to help teach and engage both veterans and civilian writers.

Thanks to Susan Solomon at *Sleet Magazine*, for featuring my writing on both military and flying topics. It's rare, in my experience, to find someone outside the veteran stovepipe who's open to submissions about those themes. She and Pamela Schmid even put me in for a Pushcart Prize for my creative nonfiction story "Chemical Warfare." They, along with Todd Pederson, are the ones who taught me you can edit poetry, too. They showed me poems aren't sacred and perfect at birth. They actually come out screaming and you have to clean them up before you show them off to the rest of the family.

Thanks to Diane Smith at *Grey Sparrow Journal*. Her encouragement and gentle suggestions have made me a better writer. And, twice, she put my poems a couple pages away from Robert Bly's poetry. Are you kidding me?!

Thanks to Lake Superior Writers for being such a supportive group. And to local literary journals that support my poetry, including *Aqueous Magazine*, *The Talking Stick*, and *The Thunderbird Review*.

Thanks to Susanne Aspley, whose novel *Granola, MN* explores our quirky place and time with grace and humor. I hope my words do similar work.

Thanks to fellow poet William Schuth, an editor at *The Deadly Writers Patrol,* for taking a chance on a fly-boy and his words this year.

Thanks to Lisa Stice, author of the collection *Uniform*, whose poems about marriage and parenting in a military family never fail to make me laugh—and occasionally wince—in self-recognition.

Thanks to editor Christopher Lyke at *Line of Advance*, a writer of searing, fearless prose. For two years running, *Line of Advance* awarded me the Col. Darron L. Wright Award for Poetry, which is very humbling and inspiring.

Thanks to Peter Molin and the extended tribe of "war-writers," some of whom also hang out with the Military Writers Guild. Molin is an essential advocate and observer of both war and literature. You can read his words at: acolytesofwar.com

Thanks to novelist Andria Williams (*The Longest Night*), for creating a network of mutual support for military voices—one that celebrates both family and veterans—through nearly everyone she meets. Citizen or service member, we're all in this together. You can read her words, and those of others, at: militaryspousebookreview.com

Thanks to poet and teacher Jason Poudrier (*Red Fields*), who engages his communities in Oklahoma in a constant spirit of healing and understanding. His example shows that a few words can inspire many.

Thanks to philosopher-comedian Steve Martin for once saying you should make room in yourself for the unexpected. I first listened to him on vinyl as a kid in my room in New Hampshire. Advice like that sticks with you.

Thanks to my grandmother, Dorothy Chandler. I dedicated this book to her because she used to bounce me on her knee saying, "Trot trot to Boston / Trot trot to Lynn / Look out! Look out! / Or you'll fall in!" as she pretended to drop me on the floor. And for reciting that Popocatepetl ditty that I wrote about in this book. Maybe I come by this (whatever "this" is) honestly. My great aunt, Alice J. Davis, wrote and published poetry. My uncle, Warner S. Chandler Jr., wrote some poems—a fact I didn't learn until I was at his funeral. Maybe there's a gene that makes you think in weird word shapes.

Thanks to my family. Thanks to my son, Sam, and my daughter, Grace. Even my dog, Leo, who doesn't usually read poetry.

And, finally, thank you to my wife, Shelley. When we dated, she snatched the Cheeto from my palm like a kung fu master and then swore at me. That's how I knew.

I should really write that story down ...

—Eric "Shmo" Chandler
Duluth, Minnesota
Summer 2017

ABOUT THE WRITER

Cross-country skier, marathon runner, and former F-16 fighter pilot Eric "Shmo" Chandler is the author of the 2013 collection of essays *Outside Duluth*, and the 2014 military-themed novella *Down In It*. His fiction, non-fiction, and poetry have appeared widely both on-line and in print.

In 2016, Chandler was the first-prize poetry recipient of the inaugural Col. Darron L. Wright Memorial Writing Award administered by the on-line literary journal Line of Advance. He repeated as the poetry winner in 2017.

He is a member of the Lake Superior Writers organization, the Outdoor Writers Association of America, and the Military Writers Guild.

A 1989 graduate of the U.S. Air Force Academy, Chandler retired after a 24-year military flying career with the U.S. Air Force and the Minnesota Air National Guard. He is a veteran with three deployments to Saudi Arabia for Operation Southern Watch, three deployments to Iraq for Operation Iraqi Freedom, and one to Afghanistan for Operation Enduring Freedom. He flew over 3,000 hours and 145 combat sorties in the F-16.

Now a commercial airline pilot, Chandler lives in Duluth, Minnesota with his wife, two children, and a dog named Leo.

ABOUT THE COVER IMAGE

October 13, 2007—A 148th Fighter Wing F-16 "Fighting Falcon" from the Minnesota Air National Guard rests awaiting the next day's inspection in Duluth, Minn. (U.S. Air Force photo by Senior Airman Donald Acton)

The Minnesota Air National Guard's 148th Fighter Wing base is located at the Duluth International Airport, adjacent to a 10,000-foot runway. More than 50 buildings are located on the 400-acre base at the end of Haines Road.

Of the state's overall strength of 2,500 Air Guard personnel, approximately 1,000 people are currently members of the Duluth unit. Of these 1,000, more than 300 people are employed at the base on a full-time basis, making it one of Duluth's largest employers.

The F-16 "Fighting Falcon" is a multirole combat aircraft, and is flown by a single pilot. (A two-seat variant is used for pilot instruction and validation.) It is highly maneuverable and has proven itself in air-to-air combat and air-to-surface attack. With a full load of internal fuel, the F-16 can withstand up to nine times the force of gravity. At altitude, the aircraft can achieve speeds in excess of Mach 2, or twice the speed of sound.

The cockpit and its bubble canopy give the pilot unobstructed forward and upward vision, and greatly improved vision over the side and to the rear. In the design, the pilot's seat-back angle was expanded from the usual 13 degrees to 30 degrees, increasing pilot comfort and gravity-force tolerance.

—*Compiled from 148th Fighter Wing press materials*

ABOUT THE COVER DESIGNER

Jeff Mlady earned a journalism degree from Drake University in Des Moines, Iowa, and has worked in nearly every facet of the publishing industry since. He writes, edits, draws, designs, and consults for a range of clients from his home in Lubbock, Texas.

DID YOU ENJOY THIS BOOK?

Tell your friends and family about it, or post your thoughts via social media sites, like Facebook and Twitter! On-line communities that serve military families, veterans, and service members are also ideal places to help spread the word about this book, and others like it!

You can also share a quick review on websites for other readers, such as Goodreads.com. Or offer a few of your impressions on bookseller websites, such as Amazon.com and BarnesandNoble.com!

Better yet, recommend the title to your favorite local librarian, poetry society or book club leader, museum gift store manager, or independent bookseller! There is nothing more powerful in business of publishing (or in poetry) than a shared review or recommendation from a friend.

We appreciate your support! In future projects, we'll continue to look for new Middle Western stories and voices to share with our readers. Keep in touch!

You can write us at:

Middle West Press LLC
P.O. Box 31099
Johnston, Iowa 50131-9428

Or visit:
www.middlewestpress.com

Made in the USA
Lexington, KY
09 November 2017